PEACE, LOVE, GOATS OF ANARCHY

Lyla

LEANNE LAURICELLA

with Alli Brydon

PEACE, LOVE, GOATS OF ANARCHY

How My Little Goats Taught Me Huge Lessons about Life

ROCK
POINT

Text © 2018 Leanne Lauricella/Goats of Anarchy
Photography on page 12 © Jana Kirn
Photography on pages 6, 8, 9, 21, 22, 106, 118, 122, 124 © Sammantha Fischer
Cover Photograph © Sammantha Fischer
All other photographs © Leanne Lauricella/Goats of Anarchy

First published in 2018 by Rock Point,
an imprint of The Quarto Group
142 West 36th Street, 4th Floor
New York, NY 10018 USA
T (212) 779-4972 F (212) 779-6058
www.QuartoKnows.com

Rock Point titles are also available at discount for retail, wholesale, promotional, and bulk purchase. For details, contact the Special Sales Manager by email at specialsales@quarto.com or by mail at The Quarto Group, Attn: Special Sales Manager, 401 Second Avenue North, Suite 310, Minneapolis, MN 55401, USA.

10 9 8 7 6 5 4 3 2 1

ISBN: 978-1-63106-565-1

Editorial Director: Rage Kindelsperger
Managing Editor: Erin Canning
Aquiring Editor: John Foster
Associate Editor: Keyla Hernández
Cover Design: Yeon Kim
Interior Design: Amy Harte for 3&Co.

Printed in China

MIX
Paper from
responsible sources
FSC® C101537

CONTENTS

Miles

Cricket

Jax and Opie enjoying their first snow together.

CHANGE

In April of 2014, my life changed forever. I can pinpoint the exact moment my existence took a 180-degree turn: the day I brought home my first two goats, Jax and Opie. I never could have imagined how intensely I would fall head-over-heels for these goats and so many others—my sweet Goats of Anarchy.

HOW DID I EVEN FIND MYSELF AT A GOAT FARM THAT DAY?

At the time, I was still a corporate gal in New York City working as an event planner. I wore stiletto heels and carried designer handbags every day, and put together extravagant events for clients with deep pockets. Think filet mignon and $400 bottles of wine. I had been doing it for years, and I was good at it. *Really* good. And I was *really* enjoying the lavish single-girl-in-the-city lifestyle: dinners out, expensive gym membership, shopping, my cute little Mercedes-Benz. There was *no way* I wanted anything to change. Then in 2011, I married my wonderful husband and we moved to suburban New Jersey.

New Jersey was not at all what I had imagined it would be. It was lush, green, and beautiful. I got hooked on the sudden change of scenery, driving around the "Garden State," looking at historic homes and farms with expansive pastures filled with horses, cows, sheep, and adorable little goats. What a change from New York City! I was intrigued.

Suddenly, my world—especially my stressful job—seemed to make less sense to me. The concrete jungle of Manhattan, the expensive lifestyle, the commute: I began to question it all. Have you ever felt like that?

For me, it was totally unexpected.

One day in my office, a coworker was talking about factory farming. I had no idea what the term meant, and for some reason I felt compelled to learn more. On my commute home that day, I searched for "factory farming" on my phone. What I saw then I can never un-see, and will never forget. The photos and videos depicted the most abject, inhumane treatment of animals. I could never have imagined this kind of thing happened. And I became *angry*. How did I not know about this? I was a born-and-bred Texas girl! Texas has farms everywhere. Back home, we would have three different kinds of meat on our plates, smothered in butter, cheese, and sour cream . . . and we loved it! I never gave a second thought to where those animal products came from. I was blind to the truth, and just followed the food traditions my family passed down for generations. Eating animals was completely normal. But that day, on my commute home, I learned the truth and was totally horrified—my entire being was shaken to the core. I knew my life would never be the same.

With tears streaming down my face, I became a vegetarian right then and there. A few months later, after more research on factory farming and those lovable, helpless animals, I transitioned to a vegan lifestyle. I had fallen madly and hopelessly in love with farm animals, and developed an intense compassion for them. Suddenly my body, my heart, and even my soul began to feel different than they had before.

Leanne and her favorite little chicken, Opal.

Ever since I was a little girl, I knew I would do something important one day. Maybe everybody has these feelings and thoughts. But I needed a serious change to get back on track toward that "something important." Once my passion for animals and their welfare was ignited, I realized what that something was. This realization, coupled with the growing feeling that being an event planner was not so important anymore, brought about a serendipitous, life-changing moment.

One Sunday, my husband and I visited a goat farm. On this fateful day, I fell hopelessly in love with goats. Their personalities were so charming! There was something so human about their faces and playful about the way they communicated. I was surprised by how engaged they were with me, and suddenly, I felt truly connected.

> **Ever since I was a little girl, I knew I would do something important one day.**

After that day, my fascination with goats grew exponentially, and I read as much as I could about them. There were so many stories about abused or neglected farm animals destined for slaughter. Did you know that most male goats born at factory farms are killed because they don't make babies or produce milk? Or that mothers have their babies taken away from them over and over to keep them lactating? The more

I learned, the more I felt an overwhelming need to run and save these little ones.

Soon my life would change even more when I brought home my first two goats: baby twins, three days old. Each weighed only a pound or two. They were the cutest things I had ever seen! They were so smart at such a young age. The twins were already interacting and communicating with each other, and with me! They recognized me and learned our routine very quickly. The goats were charming, playful, and affectionate, and had such a profound effect on my heart. They were more than just pets; they were dear companions. I created an Instagram account to share their sweetness with my friends and family.

At that time, my husband and I were binge-watching the show *Sons of Anarchy*, so we named the new goats after my two favorite characters on the show—Jax and Opie. Along that theme, I came up with my Instagram account name: Goats of Anarchy. The account became a daily document of the goats' silly, lovable antics and my new life as a wannabe farm girl.

Two months after bringing home Jax and Opie, we added three more baby goats to our family: Tig, Nero, and Otto. To my surprise, I loved the farm work and the fresh air. It felt therapeutic, and I was hooked! We now had a growing goat family, and I wanted to spend every minute with them. I was so ready for this change.

I was still commuting to New York City to a job I found stressful, frivolous, and worst of all: indoors. The more time I spent away from

Babies Tig, Otto, and Nero exploring their new home.

Otto smelling a summer flower.

Jax and Opie, the photo that Instagram featured on their home page.

shoveling poop in New Jersey, the more I dreaded going to work at an office in Manhattan. There *had* to be something more to life than waking up, commuting, working, commuting, sleeping, and then doing it all over again. I longed to do something more meaningful. There had to be a way to change my life. I stressed about it every day.

My husband had just recently changed his own life dramatically. He left a longtime Wall Street career to open a classic Corvette shop in New Jersey—his dream. It made him happier than he had ever been. This inspired me to follow my heart and do what I loved. What a tough decision to make: to give up my career and change my whole life. When I finally took the leap and told my husband I needed to leave my job and work with animals, he was fully on board. In December 2014, I traded my high heels for muck boots and my Mercedes for a pick-up truck. My Goats of Anarchy adventure had begun!

The first Monday of my "unemployment" was unsettling. I'd always had a solid job and a steady paycheck. What in the world had I done to my life? But that afternoon, I checked my Instagram account. Goats of Anarchy was gaining followers rapidly, by the thousands. Why and how was this happening? I found out Instagram had featured one of my photos on their home page and thousands of people had seen it! On the official first day of my new life, my sweet goats got over 30,000 new followers. This was a clear sign I had made the right decision to change my life.

Since then, Goats of Anarchy has grown beyond my wildest dreams. Our goats now have over half a million followers on Instagram. That's not even the best part—Goats of Anarchy has grown to become a true farm animal sanctuary and official registered 501(c)3 charity. The outpouring of love and support for us, both emotional and financial, has been overwhelming. So many of our goats have special needs and would have been euthanized or slaughtered had we not stepped in. If I had not made that one big

These little goats have changed my life and have shown me the purpose of my time here.

decision to change my life, Goats of Anarchy would not be what it is today.

This book will uncover the inside story of Goats of Anarchy. You'll learn how we began, through anecdotes about our loveable goats. Some of these stories are uplifting, and some are incredibly sad—just like life. This book also discusses the big life lessons I learned from my little goats, and you will learn how you can take these lessons out into the world. I would not be the person I am today without my goats, and neither will you. These little goats have changed my life and have shown me the purpose of my time here. They've taught me unconditional love, strength, confidence, patience, creativity, total dedication, and courage. Come along on this Goats of Anarchy journey, and you too can learn how to fight like a goat!

Phoenix

Ansel

2

FINDING PURPOSE

There is a certain aimlessness
you feel when you leave a career
behind with no real plan. I felt
both hopeful and lost, with one big
question looming above my head:
What is my purpose? I'd left my
career because I was feeling drawn to
animals. The only thing I knew
was that I wanted to work with them.

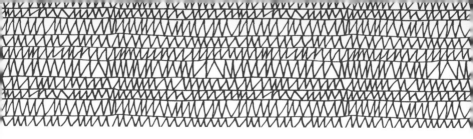

BUT ANIMAL-RELATED JOBS DON'T PAY WELL UNLESS YOU ARE A
veterinarian—and I wasn't. How else could I work with animals?

I had just left my job, but money and material things were still
very important to me. I still cared about what I looked like, how I
dressed, and what people thought of me. My focus was on how I was
going to maintain the lifestyle I was used to, instead of what would
fulfill my soul. My, how things have changed!

Feeling adrift in a sea of too-much-time-on-my-hands, I found a
farm animal sanctuary near me and decided to volunteer. I thought
this would be a great way to learn more about farm animals and
fulfill my desire to be around them. Even if it didn't help me find
my true purpose, it would keep me busy for the moment.

Me being me, it wasn't long before I fell in love with all the
animals at the sanctuary: goats, sheep, chickens, pigs, llamas, horses,
donkeys, and more! And I fell hard for two in particular: Romeo, a
mini paint horse, and his best friend Torres, a miniature donkey.
This bonded pair had been together since they were babies, and
they were just the cutest things. So I adopted them! I had to—the
sanctuary was overwhelmed and under-staffed. Now with five goats,
a horse, and a donkey, I was inching closer toward a purpose, and
a full-fledged farm.

Leanne meeting Romeo and Torres for the first time.

Early one morning, I received a breathless call from the woman who operated the animal sanctuary. She had been up all night, rescuing animals from a horrible cruelty case right in town. She told me it was a disaster. "Could you foster and bottle feed two baby goats that we rescued?" she wanted to know. That was my soft spot: baby goats. "Of course!" I said. I headed straight to the sanctuary to pick them up.

These babies were so young and tiny. And oh-my-goodness were

A very sick little Ansel when he first came to GOA.

they adorable! I named them Ansel and Petal, echoing the names Hansel and Gretel, another pair of kids who had seen horrors in their short lives.

Ansel and Petal had been brought to a livestock auction as days-old babies, and then sold to a livestock dealer. This dealer brought them back to his house in a leafy, affluent neighborhood only about fifteen minutes away from mine. It looked just like any old house in any old neighborhood. But what was happening there was horrifying. This dealer was going to livestock auctions to purchase baby animals, which he would then try to sell for profit. If he didn't sell them, he let them starve and freeze to death. Then one day, five piglets escaped from his back yard and were later caught by a neighbor who knew where they came from. When the neighbor returned the piglets, he discovered a gruesome scene and called the police.

Once the authorities arrived, they found 200 baby animals dying in cages and pens in the backyard. There was no shelter, no water, and no food. As rescuers searched for survivors, their feet crunched over bones of the already deceased. There were small animals being kept in cages on top of piles of dead animals. There were also animals being kept in the house, including Ansel and Petal. It was the tail end of winter, and freezing cold, but there was no heat in the house. And, could you even believe it—there were also two young children living among all of this. It was truly a house of horrors.

This was a big cruelty case, right in my own backyard. It was my first rescue, and it affected me like nothing else I had yet experienced—and there were my two little goats that had survived it. My heart broke for all the poor, tortured animals. How could anyone be so heartless? *How often did things like this happen?* I was starting to realize that there weren't enough people fighting for these innocent victims, and that sufficient laws to protect animals were not in place. Suddenly, I had a clearer sense of what my purpose was. I needed to save these baby goats! And not just Ansel and Petal, but also the countless other helpless animals who needed people to rescue them.

I saved these two baby goats from the house of horrors, but

Suddenly, I had a clearer sense of what my purpose was.

they were very sick. It was clear that Ansel and Petal might not survive, but I knew I'd do everything in my power to save them. My purpose for the moment was to nurse them back to health, no matter what. This was the power of my love for animals. The veterinarian confirmed they both had *E. coli*. Ansel's case was much more severe than Petal's. He could barely stand, and he couldn't eat due to extreme fatigue and nausea. The poor babies came back with me, and I set up comfy, warm beds for them inside my home. The next two weeks were spent tending to them day and night. *E. coli* is no joke and it's also contagious—so of course I got it, too! All three of us were a hot mess. I was miserably sick, but it did not deter me one bit from helping them. With new insight into how Ansel and Petal were suffering, I was more determined than ever to help them heal.

Ansel and Petal developed a following of GOA Instagram fans pretty quickly. People started checking in for updates first thing every morning, and the comments section was blowing up. Everyone was rooting for the babies to pull through. I didn't realize it then, but this community of animal lovers would become the backbone of Goats of Anarchy. Ansel and Petal's new followers fell in love with them, and were offering *me* words of encouragement too.

Ansel is feeling much better!

Ansel gives Petal a kiss.

Meeting baby Petal for the first time.

Petal regained her strength quickly and encouraged Ansel to do the same. She was his affectionate and caring goat friend, the one who had survived the house of horrors with him. But still, Ansel struggled. Even more still, his fans sent thoughts, prayers, and words of encouragement. Ansel went back and forth to the animal hospital, getting IV fluids and antibiotics. The veterinarian said there was still a good chance Ansel would not make it. He was much too young and weak. But I—having dug deeper into my purpose with animals—was having none of that! Over the next few weeks, Ansel eventually did get better. First he ate, then he walked, and then the diarrhea finally stopped. He began to act like a silly, curious baby goat. *My* silly, curious baby goat.

It was finally time to introduce Ansel and Petal to the rest of our animals. Petal was the only girl of the herd, and right away she was fearless. She let it be known that she was the queen, and she remains the queen to this day. Petal liked to be the first on the playground structure we had in the yard, and thought it was funny to jump on the donkey's back and take a ride. Ansel was a bit more cautious at first. Soon enough, he gained his confidence and we started to see a different side of him. Ansel is fiercely intelligent and boundlessly curious, and this has gotten him into some hilarious trouble! He quickly earned the name "Ansel the Destroyer." Once, Ansel ripped all the drywall out of a shed. He has also locked himself inside the chicken coop . . . more than once! Ansel regularly picks locks, breaks things, and gets into everything he isn't supposed to—just like a silly little kid. He is now the largest goat we have, solid black with very long horns. He looks majestic and intimidating but is as sweet as they come. Our sick little baby is now the king of Goats of Anarchy.

Watching Ansel and Petal grow and overcome their horrible start to life, and having a huge part in their transformation, was extremely satisfying. I was hearing from so many of our followers that our goats were going one step further than just bringing smiles to their faces—the goats were inspiring and encouraging people during rough days. Finally, I was finding my sense of purpose. The animal rescue life was for me.

Ansel and Petal were doing great, and I was ready to do it all over again.

> **Finally, I was finding my sense of purpose. The animal rescue life was for me.**

Finding Purpose

When faced with a sudden void—of direction, of money, of human company—how do you find purpose? It's so easy to seek in old familiar places, like in material things: a new car, an expensive handbag, or a fancy meal out. In my old life, I used to look to these superficial things to validate my self-worth. Or you can seek purpose in helping others, in seeing them thrive because of something you've done: a compliment, a charitable donation, or giving time to help another being who really needs it. This became the approach of my new life. Goats of Anarchy was helping me see that a worthy purpose in life was to help others: our sweet goats, and also the many followers who were looking to us for strength.

Chibs and Lyla can always be found snuggling together.

UNCONDITIONAL LOVE

After adding Ansel and Petal to
our family, my farm animal fever
continued. You could say I completely
fell in love with these animals, one
by one, as they came to Goats of
Anarchy. For several months, I was
happy and busy with my growing
farm family, posting their adorable
faces and their progress to Instagram.
We now had eight goats, several
chickens, Romeo the mini horse, and
Torres the mini donkey.

TWO NEW RESCUES ALSO CAME TO GOA: PINEY THE PIGLET AND
Prospect the baby goat. Both being babies, they lived in the house
with us and our dogs. The rest lived in our two-acre backyard.

Keeping babies in the house was a challenge. Farm animals are
not meant to be "housebroken," so I had to figure out how to diaper
Prospect. This became a messy daily experiment. How would I catch
the pee coming from his little boy parts and the poop coming from
his behind? Both could not be taken care of with a single diaper. My
solution: wrapping one around his waist and one over his bottom.
At first, we had some hilarious fails: diapers ending up as belts after
sliding up towards his middle; diapers coming off completely, with
me finding them randomly around the house; poop-filled diapers
flying across the room while Prospect energetically pranced; and
pee *everywhere*. I could never have imagined that my days would be
consumed by this! But I seriously loved these animals and couldn't
imagine it any other way.

All along, I religiously updated our Instagram account with
photos and stories of our daily sweetness and shenanigans. Every
day, Goats of Anarchy gained more followers, and I constantly
received emails and private messages from them. Some people
had questions, while others just wanted to let us know how much

Piney as a tiny piglet.

Prospect as a cute, little baby.

Twins, Chibs and Lyla, talking about goat things.

Chibs and Lyla starting their journey at GOA.

they loved our account and our mission. So many messages spoke to how these little goats impacted people's lives. Some messages were very emotional. Everyone who reached out seemed to find encouragement and inspiration from our babies. Our followers truly loved our goats, too. Their enthusiasm helped me see that I was doing something important. It provided fuel on difficult, tiring days—and there were many!

One day, I received a message about twin newborn goats in need of a special home. Both were born with congenital defects. Without

Lyla learning to balance on three legs.

giving it too much thought, I said that I would be happy to take them in. It was January and freezing outside, so of course they would have to live in the house for a while. Where else would these sweet babies go? Both were born with defects. The boy goat had contracted tendons, which meant his front legs were slightly curled and could not straighten all the way. We named him Chibs. The girl goat was born with only three legs. She was also missing her hip, and a portion of her pelvis and tail on the side with the missing leg. We named her Lyla. It was time to prepare my husband for having three baby goats, a pig, and our dogs living in the house with us, along with the rest of our farm family outside. He is such a patient

Chibs and Lyla move outside.

Chibs and Lyla enjoy their first Popsicle treat.

and understanding man, but I also think he knew he was fighting a losing battle with my growing obsession with farm animals. And, of course, we support and love each other unconditionally.

Chibs and Lyla came here when they were just days old. I fell in love with these sweet babies at first sight, and our followers did too. They both had the friendliest faces and fluffiest hair, and were snuggly and affectionate with each other. It was obvious from the start that these two goats loved each other very much. They had that "twin thing" going on and were absolutely in sync with each other. One could not be without the other, and they would not even nap unless they were curled up together. Chibs' and Lyla's infinite love for each other permeated our Instagram account.

In addition to Lyla's missing back leg, she had another defect from birth: there was a narrow tube running from her bladder to an opening on the outside of her abdomen. Urine would trickle out and run down one side of her body, causing painful urine scald. Naturally, she also had a very hard time balancing on her single back leg and was too wobbly to walk. Chibs was perfectly healthy, except for the contracted tendons in both of his front legs. Thankfully, his legs straightened on their own after a few days!

But Lyla was still facing challenges, and so was I. Trying to keep her clean and dry in the house was impossible because of this extra opening in her abdomen. There was also Prospect and Chibs to diaper. Remember how difficult it was to keep two diapers on a baby boy goat? With all three goats, the diapers would always end up around their ankles. They never stayed put. I thought, *how can I hold this all together?* Then came the brilliant idea: human baby onesies. I bought some and put them on my baby goats. They worked like a charm! Baby onesies kept the diapers in place so that my home didn't look and smell like a barn. It also happened to look freaking adorable, which was a bonus! By the way, as much as I love these goats as if they were my own babies, I don't think goats belong in a house. The only goats in my home are new orphan babies, terminal babies, and disabled babies who can't get around on their own. The end goal is almost always to move them outside to the barn and become part of the herd. That community is so important for them.

Over the next few weeks, Chibs, Lyla, Prospect, and Piney the pig all lived in the house and snuggled together by the fire. It was a freezing cold winter, so their outside time was limited. But they played a lot inside the house, as little ones would, running and prancing and gently head-butting each other. They constantly made us laugh with their antics. The goats jumped everywhere: on the bed, couch, desk, and dining table. It was always a surprise to find them in the house somewhere, after hearing the clip, clip, clip of little goat hooves on my wooden floor. They would go into the bathroom and unroll all the toilet paper. They chewed up curtains and any paper around the house. The boys had the most fun, while Lyla still struggled to walk on three legs and remained unstable while standing. Soon though she got stronger and more coordinated and was running on three legs just as fast as the boys. She was incredible! Lyla also had surgery to repair the little hole in her abdomen, and in time she was good as new.

With all the babies in the house forming their own little crew, I started to notice how gentle and nurturing Chibs was with Lyla. He would head-butt, roughhouse, and challenge Prospect, but was affectionate and patient with Lyla. When the twins snuggled together to rest, they intertwined their necks and their breathing was totally in sync. If one lost sight of the other, they called out desperately to find each other. Their love was moving. It reminded me of the unconditional love that humans feel toward their family members.

Ever since Chibs and Lyla came to GOA, I have watched their bond grow. Eventually they moved outside to be with the other goats, but they still stay near each other and stick up for each other during playful brawls. On sunny days, you can always find the two of them curled up, basking in the sun together. Before we built our new barn, Lyla slept in a stall with another special-needs goat to ensure their safety when no one was around. In the morning, we would find Chibs lying against her stall to be close to her. (Now, they have their very own stall together.) Their bond has taught me a lot about goats. Just as humans do, goats can feel unconditional love toward each other. They also grieve upon the deaths of loved ones. I sometimes worry about the day either Chibs or Lyla passes, because I don't know how the other will survive.

It's not just Chibs and Lyla who share unconditional love. We have several sets of twins, and several mothers with their babies. All the twins stick together, and all the mothers and babies stay by each other's sides for life. This doesn't just happen with goats who are blood relations. If we pair up two unrelated goats as babies, they become best friends for life, and stick together always. Goats love unconditionally, and they love hard.

Being with my goats every day, I have also fallen in love, unconditionally, with each one of them. I've become "goat mama," and feel responsible for them as if they were my own children. The goats need me, yes—but I need them too. They give me purpose, but also so much affection. They are silly and funny, and they teach me so much about life.

Unconditional Love

Think about those in your life who you love unconditionally. They could be blood relations, your husband or wife, your childhood friends, or even your own furry friends at home. Humans have an amazing capacity for unconditional love, even in the face of fear, hatred, and pain. The love inside us can grow, if we allow it. This is how we can find room in our hearts for others. After unspeakable tragedy, we can go on in part because of the immense love we hold in our hearts. I've noticed this time and again with my Goats of Anarchy. When adding a new goat to our family, or finding the strength to tackle a challenge, or overcoming grief, my capacity to love unconditionally and greatly is always what gets me through. I take this love with me wherever I go, and let it motivate and drive me to be a better person.

Angel showing off her big brown eyes and sweet little grin.

STRENGTH

Unconditional love feels amazing, but leaves you vulnerable. These goats have my heart, and I would do anything for them. Goats of Anarchy was going well, and my little farm family was growing. I needed to be strong to dedicate my life to these sweet animals. So, how to find the strength to handle the ups and downs of caring for them? And what was next for Goats of Anarchy?

I OFTEN RECEIVED MESSAGES FROM OUR INSTAGRAM followers, asking if our sanctuary was open to visitors. I hadn't thought of our operation as a sanctuary before. It was just me, a few animals in my backyard, and my smartphone—posting pictures that somehow made people think deeply about their lives. The thought of starting a legitimate rescue organization was intimidating. We had no space for that! My little backyard "farm" was now at full capacity, with no room to grow. And what did I know about starting a business, much less a 501(c)3 non-profit with strict parameters? I just couldn't do it . . . or could I?

GOA could keep going the way it was: just a few animals in a backyard. Or I could harness the strength to pursue my dream. I could toughen up and figure it out. Deep down, I knew if I wanted to keep saving these precious goats, I needed to become a legit sanctuary. You always hear stories about people coming to the United States with nothing more than a dream and a few dollars, then going on to build thriving businesses. How? By being tough, persistent, and wanting to succeed badly enough. They just figured it out.

We would surely have to move to a bigger place with more land. How would I tell my husband we needed to sell our house to buy a larger property and get MORE goats? Well, I managed to convince him! We started looking for the perfect new home for GOA and applied for non-profit status at the same time. That was it! My goats had helped me find the strength to go for what I truly wanted.

One day, a woman sent me a message about a newborn goat she found on her grandmother's farm in Kentucky. She found this baby lifeless in the snow. The woman rushed the baby into the house to warm her up, and slowly she revived. The baby goat, whom we now call Angel, was a little black and white fainting goat. Angel's mom had somehow escaped the barn and had given birth, then left her baby all alone out in the snow. When animals are born outside in below freezing temperatures, they must be cleaned and warmed immediately by their mothers to prevent frostbite. Unfortunately for Angel, her mom did not do this. She lost the tips of both ears and the bottom halves of both back legs to frostbite. We never found out why Angel's mom abandoned her, because otherwise, she is perfectly healthy.

My goats had helped me find the strength to go for what I truly wanted.

What happened next is infuriating. Angel's back leg stumps were uneven; one was longer than the other. This could have been corrected with custom prosthetic legs. But the first veterinarian who saw Angel claimed that because her stumps were different lengths, she would never be able to walk. He recommended amputation.

How is this person permitted to practice animal medicine?! This vet removed one of Angel's legs entirely and left a huge incision where her hip used to be. There was now no chance she could get a prosthetic leg on that side. Had this vet just left her alone, Angel would be walking and running today on two prosthetic back legs. He put this newborn, who had already suffered so much, through unnecessary surgery and left her cart-bound for life.

After that surgery the woman decided to reach out to me. I did not blame her for what the doctor had done to Angel. We all look to our doctors and trust that their years of education and experience will guide them. But they are human, and sometimes they are wrong. This woman loved Angel and wanted the very best for her, but she was pregnant and was unable to care for baby Angel. When she told me Angel's story, I said I would take her. I was excited to help another baby and was feeling strong and confident about my new decision to expand GOA.

Angel finally arrived, tiny and frail. She needed serious wound care for her stump and incision site, so I learned how to disinfect, medicate, and wrap frostbitten limbs and skin properly. Feeling like a good goat nurse, I could now fully enjoy Angel's company. Angel was already several weeks old when she joined us and had been lying down for most of her life. I was anxious to get her up and moving, and I had an idea. I started researching dog carts and found a company that sold something in just the right size for a tiny baby goat. I ordered Angel a little pink cart. When it arrived,

Welcoming baby Angel to GOA.

Angel getting confident in her new cart.

I was so excited. This was my first foray into carts for goats, and I thought I had totally nailed it!

I assembled the cart and gently strapped Angel in. She gave me a look that said, "What are you thinking, lady?" and didn't move. Not one step. I gave an encouraging little push, but still—nothing. I felt frustrated and defeated. But I wasn't going to give up. Over the next few days, I tinkered with the fittings on the cart. After many tweaks and tryouts, Angel started to walk. Success! She started to walk! When she began circling around my living room, it was the most beautiful thing. Angel's eyes finally said, "Wow! Thanks, Mom!" They had a new spark. She looked lively for the first time since coming home. She looked strong.

Angel and Piney both lived in the house with me and their friendship blossomed. Angel loved Piney, and Piney loved Angel. It was the most endearing thing to watch! Piney was much bigger than Angel but was very gentle with her. He would curl around Angel in what I can only describe as a "pig hug." Angel would lean on Piney and groom his hair! They would always find a way to be close, even if it looked uncomfortable. Our followers loved the two of them together. Piney and Angel helped people to see farm animals in a different light. Before our very eyes grew the strongest friendship between two different animal species. Their bond taught us all about accepting others who are different. Again, the reactions of GOA fans reinforced that what I was doing was important. Where I once thought

Angel and Piney get cozy by the fire.

I didn't have the strength to go my own way, here I was! I was doing it!

Angel showed some real progress. She got stronger and more sure-footed in her cart. But there were still challenges. Angel was

Angel being silly.

taking her bottle but had not yet started to nibble on hay, which is essential for goat nutrition. She was (and still is) very stubborn, and she absolutely refused to try eating hay. I tried giving her hay-flavored water, hay dust, grain, sweet feed . . . anything to get her to eat. Finally, I dissolved grain in warm water, kind of like a smoothie. It worked!

Once she got a real taste for it, Angel started eating more and gained weight fast. My little girl was getting big and strong. But then came problems. Angel gained weight so fast that her front legs weakened and would collapse. We had to get her a bigger cart because she could no longer fit in her little pink one. Since then, we have struggled with several carts and several pairs of custom braces for Angel, with not much success. She's even seen a physical therapist. Sometimes it seems like we take two steps forward, one step back with her. Sometimes we take no steps at all.

Through everything, Angel has been incredibly tough. She knows her limitations, but she is OK with them. She's loveable and feisty and doesn't care what she's "supposed" to be doing. She may be immobile, but Angel isn't lying in the straw, pathetic and alone. She gets 24/7 care and is always comfortable. She is kept clean, well fed, and happy, and she is in and out of her cart throughout the day. Our volunteers all have dedicated "Angel Time."

Angel has endured so much in her short life, more than any little goat should have to. You wouldn't know how she once suffered by looking at her face. Angel has taken what life has given her and has

Angel in her "big girl" cart.

adapted and embraced it. Through all of this, she has remained sweet, loving, and content. If that's not strength of character, I don't know what is!

Animals like Angel, with deformities and disabilities, are usually discarded. Watching her overcome obstacles and enjoy her life drove me to do more. It drove me to become stronger, more motivated, and more dedicated to GOA. I was sure there were other special-needs goats like Angel, and I wanted to have the opportunity to give them fighting chances, too. Shortly after Angel arrived at my little farm, GOA became an official non-profit animal sanctuary! My experience with Angel only validated my decision to take GOA to the next level.

Strength

What gives you the strength to keep going? It's hard to keep strong when life throws challenges your way. Sometimes, all you need is a new outlook to bolster your spirit and help you become courageous. When I find myself worried or weak, I think about what my goats have gone through in their short lives and I'm immediately humbled. If sweet, frostbitten baby Angel can overcome tremendous hardship, then I can buck up and get through a tough time. If my goats can find strength, then so can you. A little perspective, a positive attitude, and some serious goat snuggles can give you the power to handle any adversity.

Miles enjoying a winter walk.

5

CONFIDENCE

Even when we find the strength to
make a giant leap, it takes a little
more to sustain us after the change.
That's where confidence comes in.
Confidence can come from others
cheering you on, like our amazing
GOA followers. It can come from
categorically believing in yourself
(or faking it until you do!). Or it
can come from experience, getting
something right over and over again.

I COMMITTED TO GOA WITH FULL AWARENESS, BUT I DON'T really know how I fell into the niche of helping goats with disabilities. With no medical education or experience and having never worked with animals or humans with special needs, I didn't plan it—but somehow these sweet disabled goats found me. Though I didn't shy away from the challenge, each new special-needs goat was uncharted territory.

Like Miles. He was born with both of his front legs seriously twisted and could not straighten them at all. His back legs were perfectly straight, so he stood at an extremely awkward angle. This meant his face was very close to the ground when he walked, and his bottom lip dragged on the floor, eventually rubbing raw. Seeing him struggle was heartbreaking. It concerned me, but it didn't intimidate me. My strength and confidence came from having worked with Lyla and Angel and seeing them succeed. I accepted the challenge of helping another special-needs goat.

We went to our large animal hospital to see what could be done for Miles and soon discovered that his condition could not be corrected by splinting. He would need to have a special surgery to fix his legs if he were to have any chance of walking upright. Our vet was excited about the chance to do something a little "out of the box" for this goat. The thing is, he had not done a surgery

Miles when he first came to GOA.

quite like this before. There were two choices: take a chance on the surgery or euthanize Miles. Most people would have ended Miles's life as soon as it had begun. However, our vet was confident that he could help. We reasoned that keeping Miles in his current condition would not be a happy life for a goat. There was no question about what to do.

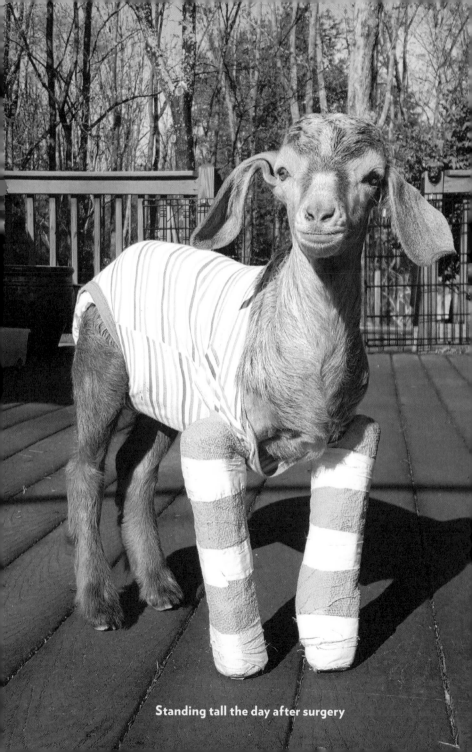

Standing tall the day after surgery

The corrective leg surgery was scheduled as soon as possible. Our vet performed a tendonectomy on Miles, which means he cut the tendons of both front legs and forced them to go straight. With both of his front legs splinted, Miles was finally able to stand up straight for the first time in his life. He held his little head high with newfound confidence. It was as if he were saying, "I can do this!" My decision absolutely had been the right one. Moments like this helped develop my confidence to keep working with special-needs goats.

Miles came back to the house to recover, and he fit right in with Angel and Piney. We had a new house crew! The three of them snuggled together by the fire and played together all day. Miles had big heavy splints on his front legs, but he was happy to finally run and play. Angel was thrilled to have a new little goat friend. Piney tolerated Miles when he jumped and climbed all over him. The three of them had the cutest relationship. After just a few days, Miles and Angel developed this wonderful love for each other. They became best goat friends.

One of my favorite things about welcoming a new baby goat to our farm is discovering what they are like. Goats are unique

Moments like this helped develop my confidence to keep working with special-needs goats.

individuals. Each looks different, and each has a one-of-a-kind personality. No two goats are the same, just like humans. Some are shy, some are feisty, some are dramatic, while others are just sweet. Miles's personality has always been very cheerful and easygoing. Mr. Happy-Go-Lucky.

All the while, my husband and I searched for a new home with a larger property. GOA was getting bigger and better, and we needed to move soon. We found the perfect place: a colonial on five-and-a-half wooded acres. This seemed like plenty of room, so we went for it and our offer got accepted. Woohoo! Goats of Anarchy was moving on up.

Now it was time to start building. Goats of Anarchy had just received our official non-profit status, and I was still thinking that our sanctuary would probably remain on the smaller side. I mean, just how many special-needs goats could there be in the world? We started to plan and build the brand-new digs for Goats of Anarchy, and it felt so right.

My goats needed a place that felt as natural to them as possible, so we made sure they could graze and climb. I envisioned a wooded goat paradise with platforms and bridges among the trees: "The Enchanted Goat Forest" is what I called it. A ton of volunteers helped us build the playground—which included a large trampoline! My vision for the next phase of Goats of Anarchy became reality. I was confident that our sanctuary could be the best place for our goats.

Best friends, Miles and Angel.

We all settled into our new home, and the animals got accustomed to their new outdoor space. Oh, my goodness, did they love that goat playground! When they first saw it, they immediately climbed all over the ramps and stairs, playfully head-butting each other. They liked hearing their hooves stomping all over, especially on the wooden deck surrounding the trampoline. That trampoline was such a new experience for them, and it was hilarious watching them discover bouncing.

Angel and Miles quickly grew, and the weather got warmer. It was time for them to make the big move outside. I love having my babies in the house, but as a goat mama, I must do what's right for them and let them explore their world. They needed to find their footing among the herd and become confident little young adult goats.

Miles's recovery from surgery had been slow, but he didn't seem to notice or care. He always had the stability of two big splints on his front legs. Eventually, Miles recovered enough to have one of the splints removed, and he showed off a perfectly straightened leg. The other leg wasn't so fortunate. It took much longer to heal and didn't grow as fast as the other leg. There was a significant length difference between his two legs. Considering what Miles started with, I was thrilled with the results. Miles is our most significant physical transformation ever at GOA, and I am proud of both him and myself. With the help of splints and custom prosthetics, Miles now stands proudly on two straight legs! I watched Miles's confidence grow as he learned to climb up on the platforms, jump on the trampoline, and play on the goat swings. Before, he would never have been able to approach the playground, but now he is the first one there every morning.

There are quite a few goats fitted with prosthetics here at Goats of Anarchy, many of which have lost limbs due to frostbite. We call

Miles growing and standing tall.

them "robo-goats," and help them to heal and walk again. When I was first faced with this challenge, I did my research and found the custom-fitted prosthetics for animals. There were plenty of dogs who had them, but not many goats, since it's uncommon for anyone to invest in something like that for an animal who is usually killed for food. But I wanted to show the world that goats are worth it, just like dogs.

Miles was one of the first, but we now have around twenty robo-goats wearing prosthetic limbs. We have great success helping them walk again, and we grow more confident each time we see another goat walk on new legs for the first time. The goats usually can't believe they're walking! The sweet looks on their faces are worth every ounce of effort. Each time another robo-goat starts to run, jump, and play with his friends, I see the life-changing miracle these custom prosthetics provide for these babies. There will always be a special place in my heart for Miles. His transformation is what truly gave me the confidence to keep taking on new special-needs goats, no matter how challenging. I will be forever grateful to him for that.

The sweet looks on their faces are worth every ounce of effort.

GOATS OF ANARCHY
Confidence

The most impressive thing about our goats is their ability to adapt to anything life throws at them, and gain confidence despite their challenges. They work with what they've got, and many people admire them for that. We can all learn from this positive attitude. Goats have a grittiness that allows them to enjoy life even if they are disabled. You can harness this confidence, too, when you might feel lacking. Just smile like Miles, who loves to show his big, beautiful bottom teeth, and think, "I can do this!" And know that the goats and I are always cheering you on.

Polly enjoying the sounds and smells of nature.

PATIENCE

With this growing confidence in myself and GOA, and now more space, we started taking on more special-needs goats, helping them rehabilitate and thrive. I found that one of the main things I needed to grasp was patience. If raising goats was challenging, raising special-needs goats was especially challenging.

SOON I RECEIVED ANOTHER PHONE CALL ABOUT A GOAT IN need of help. She was a blind baby who was living at a farm not too far from mine, and she was not being cared for properly. Of course, I was totally down to help another baby, but I had no previous experience with blind animals. *How hard could it be?* I thought.

When baby Polly came to GOA, she was shockingly small for her age, with an unusually large belly. Polly was six weeks old, but only about three or four pounds. She should have been double that. Her mother was also there but didn't pay much attention to Polly. Something just wasn't right. Poor Polly was walking around in circles, constantly crying out. She seemed confused and somewhat distressed. The people then caring for Polly worked full time and were not equipped to handle a blind newborn baby goat. The best thing to do was to get this baby home with me as soon as possible.

Polly and I had a lot to learn about living with each other. This was a first for both of us: for me, it was having a blind goat. For her, it was being in a comfortable home and having people devote time to her care. When she first arrived, I gave her the grand tour, walking her through the house slowly to "show" her around. Since she couldn't see anything, Polly needed to know where everything was so she wouldn't bump into things or get hurt. Things move

Polly wrapped in a cozy blanket.

fast at the farm, but I slowed it down for this sweet girl, and she quickly took to me as her new mama. Polly started following me everywhere by the sound of my voice.

She relied on me heavily. When she got hungry and I wasn't right there, she would pace around the house crying out. When she didn't find me right away, she would go to a corner of the room and begin sucking on the wall. I would come back in from the barn to find her in a corner with lips firmly planted. It was unsettling because Polly also seemed anxious and upset. I hated seeing her this way, and was starting to think that there was probably a little more going on with Polly than just her blindness.

Our vet confirmed Polly also had a neurological disorder present from birth. Wow. Well, that explained her anxiety and the sucking, which must have been her way of calming down during a freak-out moment. I had to figure out how best to care for Polly now that I fully knew about her condition. I knew that patience would be the key ingredient.

I knew that patience would be the key ingredient.

I have to be honest: Polly and I had a bit of a rocky start. Her anxiety started to give me anxiety, and I felt overwhelmed. I constantly worried about her. She wouldn't eat solid food, so weaning her from a bottle was impossible. Would she ever grow, move outside, and have a "normal" life like the other goats? Would she ever bond with anyone else besides me? What would freak her out and what would soothe her? We had to be careful not to move anything in our house or place objects like boxes in the way. If there was something around that she would not anticipate, she could get hurt or totally disoriented. Polly worked best on a strict routine, and new noises or voices threw her off. We had to be very careful and patient around her, always. Figuring this out was full of trial and lots of error. Even with our rocky start, she eventually learned to trust me, and I learned to accept her for the way she is. Through caring for Polly, I learned the true meaning of the word *patience*.

Still, not all was completely smooth sailing—life never truly is, right? Polly would still panic sometimes. I brought her outdoors, thinking she might fare better in the fresh air. But out in the barn, she seemed even more disoriented and anxious. Then I tried swaddling Polly with blankets, like a human baby. She loved it! A peaceful, calm look came across her face and she went right to sleep. Yes! I was excited to find that the blankets soothed her, but knew she couldn't be swaddled in blankets all day long. What to do?

Right before Halloween, I ran to the store for a few things and spotted *the cutest* duck costume for kids. I don't make a habit of dressing up our goats in costumes but thought it might look adorable on one of the goats one day. The next day, while swaddling Polly, I remembered the duck costume and thought, *maybe she'll like it*. As soon as I put the costume on Polly, she completely relaxed. She even started to drool, that's how calm the costume made her.

The legendary duck costume.

It was so cute, and like a doting goat mama, I snapped a photo or two (or 100!). The next morning, I put the costume on Polly again and she had the same reaction. This little duck costume had such a soothing effect on our anxious little Polly. I felt like I had totally scored the winning goal with this discovery. After doing a quick check on Instagram, I learned that a very special baby goat in a duck costume had broken the internet! The Instagram post got almost 22,000 likes and over 2,000 comments practically overnight. That was a lot of love for our little Polly!

Then things got even bigger. The same week that Polly and her costume were on Instagram, an online media outlet focusing on animal rights called *The Dodo* asked me to do a general interview about our goats. During the Q & A, I casually mentioned Polly and her cute little duck costume and sent them a photo of her wearing it. A couple of days later, *The Dodo* published their story featuring Polly and her duck costume. Well, their site has millions of followers and a ton of other media outlets picked up the story. The next day, I received about fifty emails from various newspapers, radio shows, television shows, online media, and print magazines from all over the world. All of them wanted to talk about Polly's duck costume and broadcast it. I couldn't believe it—Polly's story went viral!

The viral story about Polly brought much-needed exposure for our organization. It gave us momentum and brought us lots of new followers, all checking in on our little superstar. Parents of children on the autism spectrum really made a connection with Polly, and

Sweet Pepper loves exploring on her own.

related their caregiving experiences to mine. We all need to have great patience with our little loved ones.

Then one day, Pepper came into our lives. Pepper was a preemie who had been born in a complicated delivery. She may have been without oxygen for a little while—we didn't really know. When Pepper first came to GOA, she was a little tough to figure out, like Polly during her first days with us. But I had patience with Pepper. At first, we thought Pepper may be blind, but then we found out that she had a neurological disorder. What a perfect best friend for Polly! Pepper needs routine and a calm environment, just like Polly. She marches to the beat of her own drum, just like Polly. She is a happy goat, but lives in her own little Pepper-world. And she and Polly have made perfect companions who will live together for life.

Polly and I worked together to find our rhythm. But still, even the smallest things proved to be tough, and required patience from both of us. Some neurologically challenged goats are said to be "in their own world," and Polly truly lives in her own little Polly-world. She walks in circles and spins her head around. Polly took a full year to learn to eat hay, and she still refuses to drink water from a bowl. When she tries to chew her cud, she zones out just a little too much and

We all need to have great patience with our little loved ones.

Josie and Emma loved each other very much.

drools . . . a lot. Our little girl is still so tiny for her age, but she gets what she needs and finds love and happiness being snuggled by her human friends. And we love her right back.

Our GOA family has been so blessed to know several special babies, like Polly, who have neurological and physical disabilities. All of them are an inspiration!

Josie almost always holds a single strand of way in her mouth. It's kind of her thing.

There are Josie and Emma, sweet twin girls who had a special sibling bond. They came to GOA with congenital leg defects that rendered them immobile, and they also had slight neurological complications. But the girls loved life and had each other—until one day, when Emma passed away from a heart defect. Josie was distraught. I saw her cry real tears that day. Scientists say that

Scientists say that only humans can produce tears of emotion, but I personally witnessed an animal grieving. only humans can produce tears of emotion, but I personally witnessed an animal grieving. Josie and I have turned to each other for comfort many times. We cuddle together every night, and she lives in the house so she gets a lot of attention. I lovingly refer to her as my "little spoon." Josie is very loved by all of us and is a happy girl.

Cherry was a baby Boer goat who lived in what we lovingly called "Cherry's world." She liked to keep to herself and loved exploring on her own. Cherry was unique, and so special. Mothers of autistic children wrote to me because they understood "Cherry's world" and their kids related to her.

When Butterfly first came to GOA, she was walking and running in her cart, but now she cannot walk at all. Unfortunately, she has a progressive neurological condition that gets worse as she ages, and she now has limited mobility. She still seems very content and loves to gather around the basket of hay with her friends.

Finn came to us as a wobbly baby who couldn't lift his head. He has vestibular disease, which disturbs balance, so he also has a cart. It usually affects animals on one side of the brain, but in Finn's

Cherry is deeply missed, and we will always remember "Cherry's world."

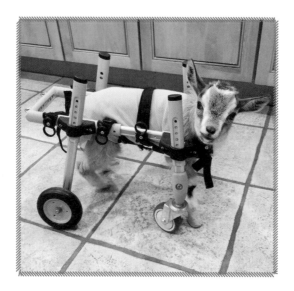

A little "Butterfly" in my kitchen.

Finn.

case, it affects both. All the GOA animals are loved, but Finn has by far stolen all our hearts. He is the first to wheel up to greet visitors at GOA and demand attention. His progress has been amazing and surprising, and I can't wait to see what the future has in store for him.

Patience

Being patient is not easy, especially with loved ones. We've all experienced moments when we wished we were more patient and calm, and regretted moments when we've blown a fuse. Most of the time, the person who tests our patience needs our patience the most. It's taken so much patience to learn the right way to care for our sweet Polly and to help her figure out her world at GOA. For me, it's been like exercising a muscle that I know will keep me strong enough to handle anything. Think of patience as something to learn and practice daily—while you're driving, at the grocery store, at home with your family, and most importantly, with yourself. If something irritates you, take a step back, take a deep breath . . . and swaddle yourself in patient thoughts, like Polly wrapped in her duck costume.

Sweet Lawson with his big soulful eyes.

GRIEF & COURAGE

On January 16, 2017, an email came
in from a woman named Chrystal—
a follower of GOA—who wrote to
comfort me after the devastating loss
of one of our baby goats, Cherry. She
also wanted to tell me her story.
All Chrystal ever wanted in life was
to be a mother. In January of 2016,
she was overjoyed to find out that
she was pregnant.

BUT IN FEBRUARY SHE FOUND A LUMP IN HER BREAST AND IN March she was diagnosed with stage 3 breast cancer. By the end of March, it was stage 4. Her doctors said she had a fighting chance with chemotherapy, and she started right away. In May, her water broke and she was put on bed rest to help stop labor. Three weeks later, on June 11, 2016, Chrystal had a beautiful baby boy she named Lawson. She was so happy to finally be a mother, even with the struggles she and her baby both faced. Lawson weighed only 1 pound and 10 ounces, but he was gaining weight and soon was up to 3 pounds. Chrystal was fighting for her own life, while her son was also fighting for his. Tragically, he took a turn for the worse and passed away in Chrystal's arms when he was only 43 days old. As you can imagine, she and Lawson's father were devastated.

At the end of her letter, she wrote that after losing Lawson she looked to Goats of Anarchy for strength, happiness, and inspiration on days when she felt devoid of joy. She thanked me for making the world a better place and encouraged me never to give up, adding a special note to comfort me over the passing of Cherry. I was astounded. This mother, who had experienced the most tragic thing a mother could ever experience, was reaching out to encourage me after losing a baby goat. I suddenly felt guilty for thinking of my goats as my kids. As devastated as I was over the

Lawson enjoying one of his daily walks in his cart.

loss of my baby Cherry, I knew it was nothing compared to the loss that Chrystal experienced.

To honor her, I decided one day we would name a special baby goat after her son, Lawson.

Well, the stars must have aligned: the very next month a message arrived from a woman about a newborn goat with a congenital deformity in both of his hind legs. Both legs seemed to be disconnected about halfway down and his legs were just kind

Lawson snuggling with his best friend, Toby.

of dangling there. The baby goat couldn't walk on all fours, so he walked on his front two legs and dragged his back legs on the ground behind him. The woman asked if I would take in this baby goat because she really didn't have the means to take care of him.

Deformities didn't scare me (still don't!), and a baby goat who couldn't walk didn't either. By this time, I had dealt with plenty of both. If a goat is not truly suffering, I will always try my best to help him or her. Disabled animals deserve a chance to live happy lives, just like disabled humans. Without hesitation, I told her I would take this baby goat. She sent me pictures and right away I knew in my gut, this was my Lawson!

"What did they do to my baby's head?!" I shouted right in the middle of my yard. My entire body filled with rage. They had TAKEN OFF HIS HORNS. This is also known as "disbudding" and it prevents the horns from growing in. People will disbud baby goats because they think the horns will injure people or other goats. To disbud a goat, a person must hold a scalding hot iron to the head of a young baby while he is restrained. The disbudding must happen before the horns start to grow in, at around one or two weeks old. I've seen it done, and I've also heard stories of blindness, facial swelling, brain damage, and even death occurring as a result. Almost all our goats have horns, and we have never had a horn-related injury of any kind. Goats have horns for a reason. They aren't just decoration. Horns are full of blood vessels to help regulate a goat's body temperature. Horns are used to explore,

He was just a baby, yet somehow also wise.

communicate, and head-butt . . . and they make excellent back scratchers!

The owner of the farm told me her veternarian talked her into disbudding Lawson. She had no idea what it entailed, the doctor told her it was necessary, and she was horrified when it happened. Here was my special baby Lawson, mutilated by a vet who thought it was necessary. Necessary? This baby goat couldn't even walk! Why put a disabled newborn goat through something like this? Who would he ever hurt? My mind was blown by the lack of compassion from a doctor who was supposed to care for animals. I now had even more fire under my feet: I was furious, but also couldn't wait to kiss that poor little burned head. I needed to get Lawson home and safe with me, as quickly as possible.

When Lawson arrived, it was love at first sight. From the moment I met my new special baby, I just felt in my gut that we were meant to be together. It was as if everything that happened since I started Goats of Anarchy was leading up to this. I welcomed Lawson with open arms, let him know I was going to be his new mama, and got him fitted for a brand-new little blue cart so he could walk around. Lawson quickly stole my heart, and also stole hearts all over the world.

This little baby goat rose up from his tragic beginnings as a force to be reckoned with! Lawson was fierce in his little blue cart

Comforting Lawson before his open heart surgery.

and nothing got him down. Every morning we went for walks down the long driveway for practice and fresh air. One moment we'd be ambling together slowly, and the next moment Lawson would take off running, his little back legs just dangling behind him! It was like he was challenging me to keep up. He was so happy, and I just loved watching him enjoy life. He was bright-eyed and curious. He was feisty and full of himself. He was a troublemaker and a bit of a bully to the other goats sometimes. Lawson had a personality way bigger than his size. He was just a baby, yet somehow also wise. Even though it might sound silly, he was my soul mate. I can't truly explain our connection, but it was something I'd never experienced with any other non-human animal. Perhaps it was that I admired Lawson and saw how much he loved life. He helped me see the way I wanted to live: with pure love and joy.

Lawson and I went to our vet to see if we could fix his back legs through surgery. The doctor gave him a thorough examination and listened closely to his heart. He listened again to Lawson's heartbeat, frowning. Something obviously was not right. As it turned out, Lawson had a significant heart defect. The vet told me there was no way he could undergo any type of surgery for his legs because putting him under anesthesia would be too risky.

I was devastated, yet determined. Lawson went to another animal hospital that had a cardiology specialist. Further testing confirmed that Lawson had a large hole in his tiny little heart, also called a VSD (ventricular septal defect). Inside Lawson's little chest, all you could hear was a swishing sound instead of a heartbeat. His condition was terminal, and I was told he may only have weeks to live. This burst a large hole right into my beating heart. My little baby! I hadn't known Lawson for long, but I have never seen a goat who loved life more than Lawson did. He was special and I was willing to fight for his life. If Lawson was sickly, weak, or unhappy, I would have considered letting him go naturally. But he wasn't. Lawson wanted to live more than anyone I've ever known.

Lawson's doctor explained the surgery that could be done to repair Lawson's heart, but we had to find a human pediatric heart surgeon to perform it. How could I even make that happen? You can't just take a goat into a human hospital and reserve a room in the OR. Feeling completely hopeless and eternally sad, I finally posted the news on Instagram.

An outpouring of love and support came right back. There were countless messages from our followers, and they all said the same thing: Lawson was an inspiration. People would say that when they had a bad day or were feeling sorry for themselves, they would just think about Lawson. I received messages from people who were battling depression, diseases, disabilities, and life struggles—things I could never imagine. Lawson kept them going. If he could live life to the fullest with the cards he had been handed, then so could they. So many people felt the same way I did about this strong baby goat, and I felt an overwhelming sense of responsibility to keep him alive.

Even with all the sympathy and inspiration pouring in, I still felt defeated. That is, until I read a message that said, "That's me, check your direct messages." One follower of Goats of Anarchy said she was a pediatric cardiologist and she wanted to give Lawson the open-heart surgery that he needed! Lawson's veterinarian was surprised when I called with this kind of news only one day after his diagnosis. Nothing like this had ever happened before!

An outpouring of love and support came right back.

A pediatric heart surgeon performing open heart surgery on a baby goat? This was the opportunity of a lifetime.

Lawson had such a fighting spirit. I saw it, and now I knew our Goats of Anarchy followers saw it. Lawson had a purpose, and I needed to keep him alive. If I didn't try, I would regret it for the rest of my life. So we got to work!

Over the next few weeks, there were lots of visits to the animal hospital, and tons of meetings, emails, and phone calls. Trying to coordinate something like this was no easy task. To our knowledge, this was something that had never been attempted before. If Lawson's surgery was successful, it could open the doors for future lifesaving surgeries for animals. The doctors told me it would all take some time to sort out, and my job was to keep Lawson healthy until then. As expected, it took a long time to coordinate the surgery, and every day felt like a hundred days. Finally, it was scheduled!

On Lawson's surgery day, I was a mess. I was not permitted to see him or be with him that day, so I laid on my couch and tried not to hyperventilate as they performed surgery on my little baby. Then I got the call. The surgery was finished, and Lawson was alive. They had done it! Lawson not only survived—he now had a perfect heart.

When I got to the hospital, I saw my brave little boy. Immediately I went to pet his head, stroking the hair between his eyes, as I always did. Then I met the surgeon who made this miracle happen.

After his surgery, Lawson had a perfect heart.

When she walked through the door, we hugged tightly and both burst into tears. She had given Lawson his life back. She had possibly developed techniques that would save the lives of many animals in the future. I'm not sure I could ever fully express my gratitude for what she did for us. She was, and will forever be, a superhero.

"Could I please borrow your stethoscope to listen to Lawson's heart?" I asked her. She smiled and said this is something the parents of her human patients asked to do. Inside Lawson's chest was no longer a swishing sound. Instead, I heard a soft, strong, healthy heartbeat. It was the most beautiful thing: the sound of a promised life.

Finally, I could sleep soundly again, my baby back at home.

The day finally came when Lawson could come home, after a long recovery. It was one of the best days of my life. We had done it! The heart surgeon, her team, and all of the doctors and nurses pulled off a miracle.

That night I slept on the couch right next to Lawson in his bed, making sure I was there if he needed me. Finally, I could sleep soundly again, my baby back at home.

The next morning, I opened my eyes and saw the sweetest face staring back at me. Lawson was awake and happily chewing his cud, just like old times. It was a moment that I will never forget.

Time to start the day! I got up, like I would any day at GOA, to change the babies' diapers and get them ready for breakfast. Lawson was looking at me, moving around and getting all excited, just like he always did. I picked him up, kissed him on his head, and put him on my lap to change his diaper.

Before I could even reach for the diaper, he let out a terrible scream. He screamed again and again, so I quickly put him back down. He was looking up at me for help and turning blue. I was home all alone and something terrible was happening to Lawson. My heart started to race, adrenaline rushed through my body, and I went numb. My mind went blank. I didn't know what was happening or what to do, and I was terrified. I called all of his doctors desperately pleading for help, but Lawson was on the floor and fading fast. Nobody was there to help him but me. I grabbed him and started giving him CPR. He started coming back around a little then, so I called our farm manager 'Arlene' who lives a mile away to come pick us up and take us to the nearest animal ER. I continued my frantic CPR. Lawson was in and out of consciousness. I picked him up to rush him to the car and made it outside but quickly dropped to my knees when I looked into his eyes. He wasn't going to make it there. I started screaming, "Lawson, come back to me! Come back to me!" But it was too late. My Lawson was gone. Fifteen minutes before, he was alert and eager to start his day with me. And now he was dead.

What had just happened? *How could I undo* what just happened?

I tried to reason. Lawson wanted to live! He had a brand-new, perfect heart! He had done great in recovery . . . how could he have died on his first day back home? My little Lawson fought until the very end, but still did not make it. The grief was unbearable, and I felt it mentally and physically.

At Goats of Anarchy, we take baby goats who are sick, dying, injured, and disabled. We do not shy away from suffering or death. I have lost quite a few babies and held many of them as they died in my arms. I am always heartbroken, but Lawson was my special one. None of the others' deaths had affected me like this.

I was in shock for days, walking like a zombie around my house, around the other animals. When I managed to announce Lawson's passing on Instagram, the response was overwhelming. Tens of thousands of condolences rushed in from people worldwide. Our followers cried, wrote me beautiful letters, and sent me paintings, cards, and photos. *New York Times* even wrote about it. But in my grief, none of it seemed to matter.

This little baby goat had affected so many people, and I had let them all down including Chrystal, who I imagined was grieving anew. I blamed myself for Lawson's death. I questioned everything, even the very existence of Goats of Anarchy.

Lost and angry, I wanted to quit GOA. The grief was too much. I could not experience the deaths of my beloved goats again and again. Lawson's death was more than I could handle, and I couldn't do it anymore. Goats of Anarchy was done. This was not the life for me anymore. I was finished.

Searching for comfort, I started spending time with the other goats. I sat quietly with Polly, Pocket, Miles, and many others. Being in their presence was healing. They were my therapy. I think I cried for about five months. Every. Single. Day. My heart is still broken and still has not healed. I don't think it ever truly will be. But the more time I spent with the other goats, the more I realized I had not been there for them as much as I should have. I had given so much to Lawson that I had almost forgotten how special each one of my animals was. What would happen to them all if I quit? And what would happen to the other babies if I quit, the babies I had not yet met? They needed me, and they needed me to be strong. For them, I had to deal with my grief and learn to carry on.

And to carry on, Lawson would be my inspiration. I would take his fighting spirit and learn a valuable lesson from his short but beautiful life. I would not quit. Instead, Goats of Anarchy would rise to the next level. Go big, or go home.

Already at maximum capacity at our sanctuary and rental property, we needed a new farm that would allow us to rescue so many more animals in need. My husband, my team, and I decided we would search for a larger farm. Raise the bar for Goats of Anarchy. Find a forever home for our sweet animals, and the many more whose lives we had not yet saved. And would you believe it? We quickly found the perfect place: a historic farm on thirty acres for all the current and future Goats of Anarchy babies to call home.

Lawson's death will never make sense. But, like Chrystal, I found the strength to keep going and became stronger in the process. Lawson made me a fighter. He gave me courage, and I will carry his heart in mine forever.

How do we find the courage to go on when faced with grief?

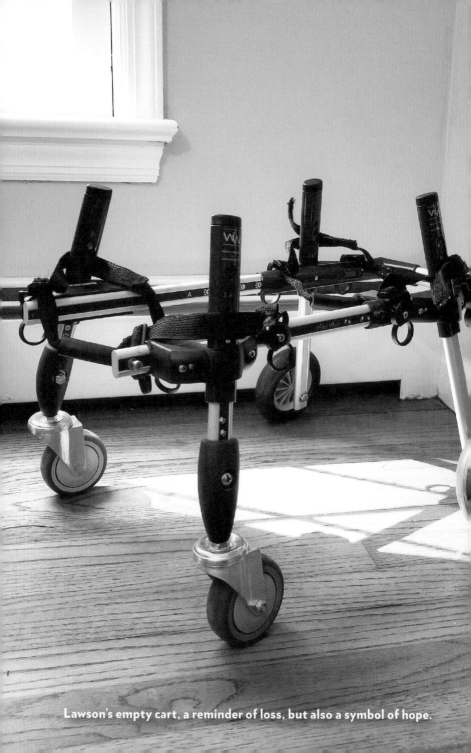

Lawson's empty cart, a reminder of loss, but also a symbol of hope.

Grief & Courage

Grief is an overwhelming emotion. It seizes every part of your body and consumes every ounce of your brain. Grief can be numbing one moment and agonizing the next. How do we find the courage to go on when faced with grief? As they say, time does heal. But time does not heal completely. To overcome the agony of Lawson's death, I looked to my goats, as well as close family and friends, for support and strength. They helped give me the courage during this difficult time. But what really uplifted me in my time of grief over Lawson's death was knowing that life has purpose. Lawson's life was full of meaning, and my life still had meaning. Realizing that alleviated the grief and helped me rise. To anyone out there who is grieving a huge loss: your life has immense purpose, your existence has meaning, and you too can rise up in the face of tragedy and find the courage to go on.

Mellie

FIGHT LIKE A GOAT

Fierce, strong, and sweet baby Mellie would become one of the biggest inspirations for me and all of Goats of Anarchy. When she came to us in early summer of 2017, she was only a couple of weeks old and so tiny— the size of a kitten! Mellie's back legs were curled up and twisted, and they dragged on the ground behind her as she hobbled on two perfectly functioning front legs.

BUT THAT LITTLE GIRL STILL WALKED WITH ALL THE VIGOR she could muster—and that's exactly how she lived, too. Mellie was a force of nature!

At the same time, we were busy planning Lawson's big surgery. Mellie's buoyant spirit matched Lawson's, and I thought the two would totally hit it off. Mellie had charmed all of us with her fighting spirit—she was a breath of fresh air at Goats of Anarchy.

Each time we receive a new baby, I do an assessment before they see our veterinarian. Within the first hour after arriving, the goats are checked and treated for lice and parasites, and are given antibiotics, probiotics, and/or pain meds, if necessary. I developed a habit of listening to the hearts of all my babies, a routine I had started with Lawson. When I placed the stethoscope against Mellie's chest, I stopped, strained my ears, and immediately knew something was wrong. My own heart sank because I knew that sound all too well: a *swish-swish* where a *thump-thump* should have been. What were the odds I would have two babies with the same heart issue at the same time?

Mellie tagged along during Lawson's next vet appointment. Lawson's doctor took Mellie back for testing. Before we left, the doctors asked to speak with me in private. I could just tell they did not have good news. They confirmed that Mellie did have the

Mellie already a pro in her first cart.

same exact heart condition that Lawson had, a VSD (ventricular septal Defect).

At the same time we were planning for Lawson's once-in-a-lifetime surgery, there was a baby girl who would never have that same opportunity. Lawson's surgery would be a miracle. There'd be no way we could ever do the same for sweet Mellie. Both baby goats were like children to me. How could I give one of them an opportunity to live, while the other would surely have to die? It felt like being ripped in two.

Of course, Lawson passed away soon after his surgery, which was shortly after Mellie joined us. So, these two special goats didn't know each other for long. But they were so much alike. Where Lawson had spunk, so did Mellie. She was full of life . . . and attitude. That girl was sassy! Mellie was easily the smallest goat at GOA, but her personality was huge. Just like Lawson, she was a beast in her little cart. My tough girl. No sooner would I place her in the cart, then she would take off at full speed around the house. I could barely keep up! Part of me was thrilled she was enjoying life, and part of me was frightened every time I saw her being "reckless" in her cart. Letting her zoom around like that was a risk.

But what's life, really, without a little risk? What's the purpose of life without joy and excitement?

So, I decided I would let Mellie live her life, rather than trying to keep her still and quiet to preserve her for as long as possible. She would stay in the house with me, and we would make sure

Mellie's first Popsicle.

she enjoyed every minute she had left. We started "Mellie's Bucket List"—a list of amazing, fun, delicious things for her to do before she eventually passed away. I posted about it on social media so our followers could also celebrate her life with us. GOA fans loved participating and chimed in with their own bucket list ideas for Mellie. It helped us all accept the inevitable and also enjoy watching Mellie live and love her life. Some of the things that were on her bucket list:

- Play in the snow for the first time.

- Swing in her very own custom-made swing.

- Eat a Popsicle (or ten!).

- Try new treats and discover her favorite . . . tortilla chips!

- Try out the trampoline.

- Meet the "big kids" on the farm.

Mellie checking out the "big kids" playground.

Every morning, the house routine is this: feed the babies their bottles, change their diapers and onesies, and put them in their carts. Then the baby goats all gather around a basket of hay for breakfast in my "barn office." Mellie added her own twist on our morning routine. She would take off at full speed in her cart and loop circles around the first floor of the house several times, then settle in front of the hay to eat with the others. She just had to get her morning exercise! Over several weeks and months, these laps around the house became fewer and slower. Eventually, they stopped almost completely. Mellie's breath seemed labored and she would get tired easily. My spunky girl was slowing down, but she didn't lose her spark. She really would push herself to her limits. While Mellie continued to fight, I was determined to fight right alongside her—for as long as she wanted to keep going.

Our followers watched Mellie and were inspired by her energy and her will to live. Anybody could see she was vivacious—just like Lawson, she fought to live on her own terms. Awestruck by Mellie's spirit, we started saying the phrase "Fight Like a Goat" around the farm, and it caught on. I added the hashtag #fightlikeagoat to every post about Mellie. People related to the phrase, which is really about staying tough to get yourself through hard times. It's about perseverance, and sometimes also about perspective. When people tell me they look to our babies for inspiration when life gets rough, I tell them to stay strong and fight like a goat. Because I've never seen anyone fight as hard as our little goats do.

Awestruck by Mellie's spirit, we started saying "fight like a goat."

Eventually, Mellie's activity slowed to a crawl. Then she could no longer walk in her cart, too winded whenever she exerted energy to move. Her little heart sounded like it was about to swish right out of her chest. She didn't show any pain outwardly, but she was slowly declining. At times, Mellie had trouble catching

Even while immobile and physically weak, she was tough.

her breath after something as simple as taking a bottle. She couldn't get in her cart anymore so instead of putting her in there for breakfast, I started placing her in a laundry basket beside her little goat friends' where she happily munched on hay. But in true Mellie style, she would often get feisty and start head-butting her laundry basket as a threat to whoever was closest to her food. Even while immobile and physically weak, she was tough.

The day would come when we would have to say goodbye to Mellie, and I had a feeling it would come soon. From past experiences, I also knew her death would probably be a painful and terrifying one. It would likely be a heart attack, not drifting off quietly in her sleep. I had watched Lawson die, my sweet baby in pain and terror, for the longest fifteen minutes in the world. He really struggled, and he was scared and confused. His last moments on this earth were torture. So, as Mellie declined in health, I made the gut-wrenching decision not to let her go through that. Our vet would euthanize her when the time was right.

One Saturday morning in early 2018, I picked up Mellie to change her diaper and noticed her struggle to catch her breath. Like, really struggling—more than I had ever seen. Her heart was racing, and she seemed lethargic and very uncomfortable. The look

Mellie in her party dress.

in her eyes said that her spark was quickly fading. She had fought hard and long enough. I knew what I had to do. I called my vet to arrange an appointment for later that day, giving us a few hours to prepare and spend some beautiful last moments together.

Mellie's favorite people came over to GOA to hug her and say goodbye. We chatted about Mellie's bucket list, from which we had ticked off some pretty amazing things for a goat. Our sweet girl had had an extraordinary life. It was time.

The vet came to our house, so Mellie could be in true comfort for her final moments. Her favorite people stayed on, and they each took turns holding her in their laps, stroking her face as she closed her eyes. Mellie enjoyed all of the attention, melting into the arms

Mellie in her final days. She is deeply missed.

of those who chose to hold her for the last time. She seemed totally at peace and was surrounded by a group of people who loved her very much. Then the first needle went in. We gathered around Mellie and gently placed our hands on her, letting her know she was not alone and was very loved. A minute later, the second needle went in and she was gone. It was devastating and beautiful at the same time. Letting her go in peace and comfort was the greatest gift to her. It was one of the hardest things I ever had to do; I loved Mellie as if she were my child. But when I needed the strength to let her go, she gave it to me. Mellie had really taught me how to fight like a goat.

Fight Like a Goat

I've seen firsthand how life can knock you down. Challenges and hardships can be devastating and feel impossible to overcome. We all go through these times. Letting Mellie go was, for me, one of those times. But if Mellie could be tough, then so could I. If she could find courage to live life with ferocity, then I could find the courage to go on without her. And so can you. If you take any inspiration from our brave goats, it would be to fight like a goat. Fight with all your might, and for as long as you can. Just keep going. You'll come out the other side and have the courage to do anything.

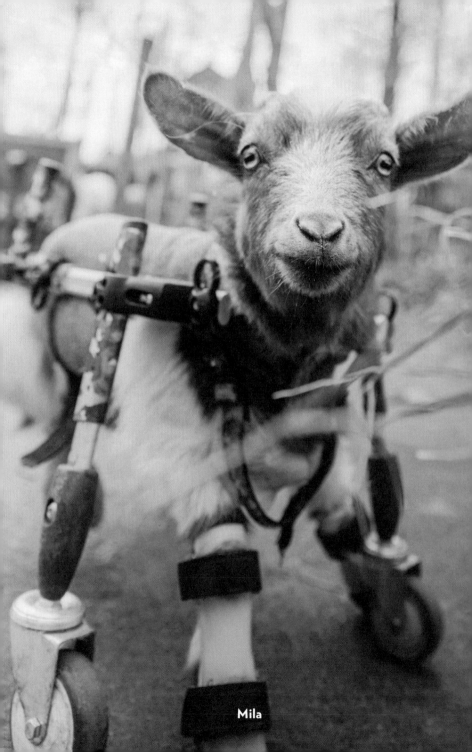
Mila

HOPE

When I walk outside and look around at all the adorable, little faces of Goats of Anarchy, I am overwhelmed with accomplishment and pride. A few years ago, when I only had Jax and Opie, I never imagined what GOA would one day become. Before that, when I was running myself into the ground doing a job I found meaningless, I *really* could never have imagined what my life would one day become. It blows my mind.

GOATS OF ANARCHY IS NOW AN OFFICIAL NON-PROFIT farm animal sanctuary, improving the lives of animals every day. Yes, we have created a safe place for animals, but we've also fostered a strong community for the people whom we lovingly refer to as our "GOA Army." Our goats inspire, and touch, and sometimes even save people's lives just by being who they are. I watch with wonder as our worldwide followers check our daily posts and leave comments, questions, and supportive words for us and each other. They educate and encourage one another under our posts. Hundreds of thousands of people gather together in an online space to check in on our furry farm family. There is so much love here, and so much hope. So many people have told me that our goats have encouraged them, inspired them, and even lifted them out of the depths of despair. We aren't just helping goats—we're helping people, too. It is humbling, thrilling, and terrifying all at once.

And the future of Goats of Anarchy is so exciting! We've finally found our forever home: a beautiful historic farm on thirty acres with large pastures, a pond, caretaker housing, and a massive barn to have all our goats under one roof. We have a veterinarian office, quarantine stalls, event space, and plenty of room to expand in the coming years. This means we can run exactly as we want to, in one large location, with all the staff and animals coexisting in an impactful way. It's more than I ever could have hoped for.

This move held so much emotion for me. My husband and I loved our personal home and the goat paradise that we created around it, but if we stayed, it meant that GOA could not save more lives. So, in the interest of the mission of GOA, we gave up our home and moved to our new location, along with other staff members who had also made the same decision. At first, I was nervous to move to a larger property with more responsibility, but by moving, we are saying GOA isn't going anywhere. We are here to stay, and we are just getting started! Having a place where all our animals, employees, and volunteers can be together was my dream. Now, Goats of Anarchy feels more like a sanctuary than it ever has, a place where great work can be done for years to come. It's a place that truly belongs to the animals.

> **Now, Goats of Anarchy feels more like a sanctuary than it ever has.**

While we left behind many memories from our former locations—both happy and devastating ones—we are looking to the future of GOA with hope in our hearts.

Only a few years ago, I was unsatisfied with my life and hoped for something more fulfilling. I took a chance and risked

Opie

This life hasn't been easy, but it is so worth it.

everything. I am so happy that I took that leap of faith and went for it. Never settle for a life that doesn't make you happy. This life hasn't been easy, but it is so worth it. I've met some of my closest friends through GOA. We have the most amazing staff, a trustworthy board of directors, and awesome volunteers. I consult daily with two other rescue organizations: Rancho Relaxo, founded by Caitlin Cimini, and Twist of Fate Farm and Sanctuary, founded by Ashley DeFelice. We bounce ideas around and support each other during times of struggle. It's amazing to have input and support from others who are going through the same things. Derrick Campana of Bionic Pets makes all the prosthetics for our robo-goats. Dr. Christina Wilson is our farm veterinarian who visits our sanctuary weekly. Dr. Arlen Wilbers at Quakertown Veterinary Clinic performs all our surgeries and oversees hospital stays. Over the last couple of years, I have developed close relationships with these great people as we work together to save the lives of our precious goats.

My new friends are true friends, and like-minded people. We are joined together by our love of animals, and we believe that animals are on this earth *with* us, not *for* us. Every animal's life is to be treasured, and that is what we will continue to teach at Goats of Anarchy. Being part of a close-knit family like GOA has given

June

me strength when I needed it most, and keeps me hopeful for the future of the organization.

And the goats. Our baby goats, whom I love like they are my own children. Lawson, Polly, Ansel, Petal, Angel, Mellie, and of the rest of the goats—they've taught me lessons I needed to learn. They've made me more human, compassionate, motivated, and creative. They give me hope and pave the way for all the future goats of GOA. I look forward to spending the rest of my life learning from them and loving them, too.

GOATS OF ANARCHY

Hope

Hope is a feeling of future success that's built on a strong foundation. Hope also grows from pride in your accomplishments and recognizing the unique path you've paved to get where you are today. Goats of Anarchy is my foundation, and you have yours, too. My goats, family, colleagues, and followers are my support network. What is yours? Look inside to find your strength and all around you to find your support, and you will also find hope.

Now, go out there and fight like a goat!

IN EARLY SUMMER 2018, WE MOVED OVER 100 ANIMALS to our new farm in one day! Our amazing volunteers let us fill their cars with goats, and we all followed each other in one long caravan back and forth all day long. Some of the goats even road shotgun! We now have over 80 goats, 8 sheep, 6 horses, 3 mini horses, 1 mini donkey, 3 alpacas, 2 pigs, chickens and dogs all living happily together at one location. For the first couple of days, we let everyone get used to their new home and their assigned stalls. Soon after that, we let a few goats at a time explore their pastures and practice finding their new stalls for bedtime. We were a little nervous introducing the two herds together, but one day we just went for it! It was a total success. Everyone got along perfectly! Our new property is everything I dreamed it would be. I can't wait to see what the future holds for GOA at our new home.

Finally, every GOA goat can sleep safely together under one roof.

Inside the dream barn, there is plenty of room to eat, sleep, and play.